Science Experiments

BUOYANCY

by
John Farndon

BENCHMARK BOOKS

MARSHALL CAVENDISH
NEW YORK

Marshall Cavendish Corporation

99 White Plains Road

Tarrytown, New York 10591

© Marshall Cavendish Corporation, 2003

Created by Brown Partworks Limited

Library of Congress Cataloging-in-Publication Data

Farndon, John.
Buoyancy / by John Farndon.
p. cm. – (Science experiments)
Includes index.
    Summary: Presents information on floating and sinking in liquids or air, providing
instructions for relevant scientific experiments.
    Contents: Floating and sinking – Floaters and sinkers – Why things float –
Measuring buoyancy – How ships float – Floating steadily – How fish swim – Floating
liquids – Hovercrafts and hydrofoils – Making a hovercraft – Submarines and
submersibles – Floating rocks – Floating in the air – Experiments in science.

ISBN 0-7614-1467-3
    1.Floating bodies—Juvenile literature. 2. Buoyant ascent (Hydrodynamics)—Juvenile
        literature. [1. Floating bodies—Experiments. 2. Water—Experiments. 3.
        Experiments.] I. Title. II. Series

QC147.5 .F37 2002
532'.25—dc21   2002019571

Printed in Hong Kong

PHOTOGRAPHIC CREDITS

t – top; b – bottom; c – center; l – left; r – right

Ardea: p16 Kev Deacon
Image Bank: p8 G. K & Vikki Hart; p24 Cousteau Society; p25 Jeff Hunter
Mary Evans: p4
NASA: p26 GSFC/JPL MISR Science Team; p29
NHPA: p5 Rod Planck; p17 Anthony Bannister
Pictor International: p9; p12; p14; p20; p28
Regatta: p10
Science & Society Picture Library: p21 Science Museum

Step-by-step photography throughout: Martin Norris

Front cover: Martin Norris

Contents

FLOATING AND SINKING

Throw a stone in a pond and it vanishes beneath the surface instantly, then plummets quickly to the bottom. Throw a cork into the pond and it bobs on the surface.

It may seem at first sight that the stone sinks because it is heavy and the cork floats because it is light. This is not quite true. A cork that was so big that it weighed just as much as the stone would still float, and a stone that was so small that it was as light as the cork would sink too.

Whether something floats or sinks depends not on its weight but on its density. Density is the amount of weight that is packed into a particular space. A stone is said to be dense because a lot of weight is packed into a small space. A cork has a low density because very little weight is packed into the same space.

The density of water lies in between that of stone and cork. The stone sinks because it is denser than water, and so feels the downward pull of Earth's gravity more. The cork floats

In the real world

The first boats made by prehistoric humans were hollowed-out logs.

THE FIRST BOATS

Cave paintings show that Aborigines arrived by boats in Australia 50,000 years ago. None of their boats survive, but many early boats were hollowed-out logs, called dugouts, which float because wood is less dense than water. The islands of the Pacific may also have been colonized by people in boats like these. The oldest known remains of a boat, however, are of a canoe, dating from 8,000 years ago, found in Pesse in Holland. This was made by stretching animal skins over a frame of bent wood, making a very light, maneuverable boat.

Did you know?

Humans generally float because, although their bodies are 70 percent water, the rest is tissue, which is less dense than water. But there is so little difference that a swimmer must spread his weight over a wide area for the water to bear him up. A high-diver focuses his weight to plunge right in.

because it is less dense than water, and so feels the pull of gravity less.

Things will sink into any liquid, not just water, if they are denser than the liquid. They will float if they are less dense. The same is true of gases such as air, too. Things float on the air if they are less dense than air but fall if more dense.

Most hard solids, however, are too rigid to allow even dense substances to sink. So even though less dense substances are supported, they cannot be said to be floating. However, dense things will sink into soft solids, just as an automobile sinks into soft mud. Given plenty of time, dense things may even sink into quite hard solids.

Water density is used as a standard by scientists. Every cubic centimeter (cm³) of water weighs exactly 1 g at 39°F (4°C). Things that weigh more per cm³ will sink; those that weigh less will not. Oak wood has a density of 0.6 to 0.9 g/cm³, so it floats. The metal lead has a density of 11.3 g/cm³, so it sinks.

Although they are rooted in the pond bed, water lilies float because they have air pockets in their stems and leaves.

FLOATERS AND SINKERS

You will need

✓ Household objects for testing. Choose materials that will not be damaged by water

✓ Wooden and metal spoons

✓ Various fruits and vegetables including oranges

✓ Gravel and pumice stone

1 Fill a basin with water. Test if similar things like spoons always float, or does the material they are made of matter?

What is happening?

The main thing this experiment shows is that some materials sink and others float. If you lift the materials that float, such as wood, you will find that they all feel lighter than those that sink. They float because they are less dense than the water. Materials like metal and stone usually feel heavy and sink because they are more dense. The pumice stone is an exception because it contains so many air bubbles that it is less dense than water, so floats. The orange floats because the peel contains air bubbles, but when it is peeled the air is lost and the orange sinks.

2 Do the same things always float? Or can altering them make them sink? What happens if you peel an orange?

Put a fresh egg in water and it sinks, but what happens with a very old egg? If it has gone bad, it may float because gases form inside it. Crack an egg in water to poach it and you should find that it sinks slowly to the bottom. But what happens if you add salt to the water? Adding salt makes the water denser, so will the egg now sink or float in salt water?

Test all your objects and you will find some things float and others sink. You may also find that some things, such as a washcloth, float when dry then sink when they become full of water. Other things that float at first, like the orange, will sink when they are altered in some way. See if you can alter any sinking things to make them float. Make one list of those that sink and another of those that float. Then see if you can work out what the things that float all have in common.

WHY THINGS FLOAT

It is not what is above the surface that keeps things afloat, but what is below. The duck is supported in the water by the weight of water pushed out of the way by the portion of its body underwater.

Did you know?

Ducks use air to keep them afloat. They have light, hollow bones and little air sacs in their bodies that work like the floaties children use in swimming pools. They also have special glands that secrete a waxy liquid, which makes their feathers waterproof and so trap air.

In focus

BUOYANCY

Whenever something is floating in water, it moves water out of the way—and the water pushed out of the way pushes back with equal force. This force, or buoyancy, is not just theoretical; things actually weigh less in water. The more water they displace, the greater the weight loss. This is why in a swimming pool, a child can often lift a full grown adult for a moment—something they could never do on land. Astronauts take advantage of this by training for the weightlessness of space immersed in deep tanks. Similarly, people with bone and muscle problems are often treated in pools to relieve the stress imposed by the body's own weight on land.

Water makes people so light they can move easily in water, with every bit of the body supported.

It is easy to discover that things denser than water sink, while those less dense float, but it took the genius of the great Greek scientist Archimedes (287–212 B.C.) to work out why.

Whenever something such as a stone sinks, it moves water out of the way. Scientists say it "displaces" water. Archimedes realized that the amount of water displaced when a stone sinks is the same volume as the stone. The stone takes the water's place as it sinks.

When an object such as a ball floats, it rarely sits right on the surface. Instead, it sinks a little way into the water. Like a sunken stone, it displaces water. The floating ball only displaces a small proportion of its volume, but this displaced water is the reason it floats.

Archimedes realized that the displaced water weighs the same as the floating object. Indeed, a floating object sinks into water until it has displaced its own weight in water and no further.

The object floats because as it sinks, its weight pushes water out of the way—but the water around pushes back with a force equal to the weight of water displaced. So the water gives an upward push or "upthrust." The object will sink until its weight is equaled by the upthrust of the water, and then sink no more. So a floating object is not simply sitting there, it is held by a force. This is called buoyancy.

MEASURING BUOYANCY

You will need

- ✔ A tall glass pitcher
- ✔ A thin plastic beaker
- ✔ A large, shallow dish
- ✔ An apple
- ✔ A pen for marking glass

1 Mark the water level in a pitcher of water, then drop an apple in. Mark the new water level. You will see it is higher.

In the real world

This girl is wearing an inflatable PFD that keeps her floating face up in rough waters even when wearing bulky clothes. The metal tube is the gas cylinder.

LIFESAVERS

For anyone thrown into water by accident, extra buoyancy can be a lifesaver. For watersports, people use buoyancy aids—compact collars that help swimmers stay afloat in calm water. But for non-swimmers—and swimmers in rough waters—a bulkier life jacket or personal flotation device (PFD) is vital to keep them floating face up. Some PFDs are already buoyant because they are filled with light plastic foam or kapok fiber. Others, like those on aircraft, are normally flat but inflate instantly as compressed carbon dioxide gas is released from a canister. Some inflate automatically, others when a ripcord is pulled.

What is happening?

The apple is kept afloat by the upthrust created as it pushes water out of the way. This displaced water is caught in the beaker. The water in the beaker is the same volume as the submerged part of the apple, yet it is heavier because water is denser than apple. In fact, the water in the beaker weighs the same as the whole apple. The apple has sunk until its weight is balanced by that of the water displaced.

2 Take the apple out. Place the pitcher in a dish. Fill the glass slowly with water until it just overflows into the dish.

Place a beaker right under the spout of the pitcher. Then, with your other hand, lower the apple into the water so that the water spills out into the beaker. Make sure you catch every drop that spills. Then let go of the apple altogether. Now take it out. Hold the apple in one hand and the beaker of water in the other and compare their weights. They should weigh the same. You can confirm this fact exactly with a kitchen scale.

HOW SHIPS FLOAT

Ocean liners show just how big and heavy ships can be yet still float. The **Queen Mary,** *launched in 1934, is still one of the largest ships ever built. It weighs more than 80,000 tons (72,560 tonnes) and is 1,019.5 ft (more than 310 m) long.*

Wood, cork, and apples float because they are made of materials that are less dense than water. A ship can be built of much heavier—and stronger—materials, like steel, yet still float. The key lies in the air trapped inside the hull, or body, of a ship.

Steel is a very dense material. In a ship it is shaped into a thin plate, so that a ship is made mostly of air, not steel. Whether or not a ship will float depends on the average density of its hull, which is quite low because of the air. So as the ship sinks into the water, its air-filled hull displaces a huge amount of water—exactly enough to create the upthrust to support the heavy steel.

Did you know?

The air inside steel hulls allows water to support very heavy ships. The heaviest ship ever built is the oil tanker *SeaWise Giant*, launched in 1976, which weighs more than 645,000 tons (585,015 tonnes) when laden. Plans are afoot to build a huge floating hotel called the *Freedom Ship*. If built it will weigh 3 million tons (2.7 million tonnes) and be well over a mile long.

No matter how heavy, a ship will float providing it contains enough air to keep its average density lower than that of water. So ships can contain hundreds of thousands of tons of steel yet still float. It is the weight of water displaced that keeps a ship afloat. When experts talk of a ship's weight, they refer to its displacement—that is, the weight of water it displaces.

Because it relies on the air in the hull to keep it afloat, a ship will sink quickly if a hole allows water in, or if it tips over so that water floods over the top. To keep it upright, a ship is made bottom-heavy, and also filled with a "ballast" of heavy sand to weigh the bottom of the ship down and stop it toppling over.

In the real world

THE PLIMSOLL LINE

Water is not always the same density, so ships float at different heights according to the water. Ships float higher in salt water, for instance, than in freshwater, because salt makes water denser. They also float higher in dense, cold winter seas than in lighter, warm tropical seas. So a ship loaded to the maximum in cold sea water might sink dangerously low if it enters warm river water. This is why most ships are marked on the hull with a scale called the Plimsoll line. This line shows the maximum levels the ship can safely be loaded to in various types of water.

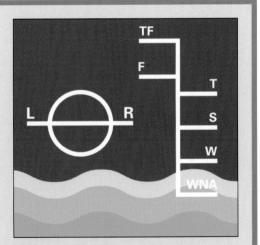

By law, every ship must carry the load lines championed by British MP Samuel Plimsoll in the 1870s.

FLOATING STEADILY

You will need

- ✔ Two plastic soft drinks bottles with screwable caps
- ✔ Plasticine or re-usable poster putty
- ✔ Scissors
- ✔ Basin or tank of water

1 Cut the bottle in half, stick a blob of putty on the rim, and try to float it. Since it is top-heavy, it falls over and may sink.

In the real world

For serious canoeists, there is nothing to match the thrill of riding a canoe through rough, white water on a mountain stream— but they must have a good roll technique to right the canoe should it tip over.

CANOE ROLL

Canoes are very light, with much of the weight quite high up. Often used in rough waters, they frequently tip over. Once over, the canoeist's weight pulls him down so that he hangs upside down underwater. The first lesson many canoeists learn is how to right the boat. There are various techniques, but the aim is always for the canoeist to shift his weight so that it pulls him back up, like the weighted bottle in the experiment. In a simple "roll," the canoeist shifts his upper body and one knee, so his weight pulls the boat back, while he uses his paddle to push up his body.

What is happening?

Boats need to be stable to stay afloat. With the weight at the top of an open boat (the half bottle), it can easily capsize, flood with water, and sink. Adding weight to the bottom makes a boat less likely to capsize and sink. The weight can be added on the outside as a keel, or the inside as a ballast of sand—here, the putty on the bottle. This is how most big ships work. Additional fins on the hull stabilize it still further. Many rescue lifeboats combine the weight in the bottom with a watertight hull to make a self-righting, virtually unsinkable boat, as in the uncut bottle-boat.

2 Stick the blob of putty on the base of the bottle half. It should now float upright quite stably.

Stick a ridge of putty down one side of the uncut bottle and screw the cap on. Now, no matter how you try to sink or capsize it, it always rights itself.

HOW FISH SWIM

Many creatures such as fish, whales, mollusks, and insects live underwater. To move around easily in the water, they need to be buoyant enough to float, yet not so buoyant that they keep bobbing up to the surface all the time.

To keep them afloat, many fish have a special gas-filled sac in the top of their bodies. Without this air sac, or "swim bladder," the fish would have to keep swimming all the time to avoid sinking.

Just as the gas in a life jacket helps a person stay afloat, so the gas in the swim bladder helps a a fish stay "neutrally buoyant"— neither floating nor sinking.

As a fish swims deeper, the extra pressure of water squeezes the gas in the bladder, so that the bladder gives less buoyancy. To avoid sinking, the fish inflates the bladder with extra gas made in its blood.

When the fish swims higher again, the water pressure is reduced. So the extra gas is absorbed back into the fish's blood and bubbled out into the water through its gills. All these adjustments are made automatically by the fish's nervous system.

Unlike other fish, sharks and rays have no swim bladder to help them float. Instead, they rely on their large, oil-filled livers. Oils and fats are usually lighter than water, so give the shark extra buoyancy. Even so, sharks always drift down to the seabed whenever they stop swimming for long.

All fish have their own personal buoyancy aid— a gas-filled bladder that allows them to stay neutrally buoyant at any depth.

Did you know?

Some catfish use their swim bladders to make noises, just like a set of bagpipes, while other fish use their bladders to make sounds to communicate. Lungfish of tropical rivers use their bladders as lungs when stranded on mud in the dry season.

Whales do not have a swim bladder. Their bodies are enveloped in a thick layer of blubber. This gives them extra buoyancy, as well as helping to keep them warm and providing an energy store.

Underwater insects are so light they do not need to be quite as buoyant as fish. Even so, many create or trap bubbles of air or gas around them. These provide buoyancy, as well as a mobile air supply. The water spider also lives

In focus

JELLYFISH

The bodies of many surface-living jellyfishes are made up of around 97 percent water, so they float naturally near the water's surface. Jellyfishes can move through the water by flexing their bendy, bell-shaped bodies but most are not strong swimmers. Jellyfishes also have basic senses that tell them the difference between light and dark. This helps them find their way.

The Portuguese man-of-war is topped by a large gas-filled float that keeps it at the surface of the sea and supports trailing tentacles up to 70 ft (21 m) long.

permanently in a bubble of air underwater, from which it launches its deadly attacks on passing prey.

Water plants also need buoyancy aids. Nearly all water plants have special air passages in their stems and roots.

In emergent plants—plants rooted in shallow water—these air passages not only carry oxygen from the leaves to the waterlogged roots, they also help keep the plants floating upright in the water. In fully submerged plants, the air passages provide essential stores of oxygen too.

FLOATING LIQUIDS

You will need

- ✔ Rubbing alcohol
- ✔ Assorted food colors
- ✔ Cooking oil
- ✔ A tall glass
- ✔ A pitcher
- ✔ Water

1 Add a few drops of food color to water and pour the water into a straight-sided glass to about a quarter full.

In focus

DEEP OCEAN CURRENTS

The surface of the ocean is always on the move, blown by the wind. The whole ocean is on the move very slowly, too, driven by differences in the density of seawater. These differences are partly created by temperature—cold water is denser than warm—and partly by the amount of salt. Typically, dense water forms in polar regions where it is cold and weighed down by salt left behind when the sea freezes to make sea ice. This dense polar water sinks and spreads deep down, outward toward the Equator. Warmer water floats back over the top toward the polar regions.

Warm water spreads from the Equator near the surface

Arctic ice

Antarctic ice

Cold, dense water sinks and moves toward the Equator deep down

Once the oil has settled into a clear layer, add food color to the rubbing alcohol. Tilt the glass gently, as before, and pour the alcohol down the side of the glass so that it flows across the top of the oil. Take care not to pour so fast that the alcohol breaks through the oil. Stand the glass upright.

2 Tilt the glass and slowly pour the oil so that it flows over the water without mixing. If it mixes, leave it to stand.

Rubbing alcohol

Oil

Water

What is happening?

Just like solids, liquids that are less dense than water float, providing they do not mix. Oil is less dense than water. It does not mix with water, so it floats in a layer on top of the water. This is why you can often see multicolored patches of oil dropped by vehicles floating on puddles on the road. Rubbing alcohol is less dense than either oil or water, so it floats on oil. However, rubbing alcohol mixes easily with water, so if it was not for the layer of oil, it would mix in with the water and so become indistinguishable from it.

HOVERCRAFTS AND HYDROFOILS

The world's largest hovercraft weigh 150 tons (136 tonnes) and ferry cars and people across the English Channel.

By sinking right into the water, ships get ample support, but in order to move they have to push a lot of water out of the way. This is why ships almost always move slowly. Hovercraft and hydrofoils, however, skim lightly across the surface and so can go much faster than ships can.

Did you know?

French railroads are developing a train that will run along a single monorail track on a cushion of air just 0.1 in (25 mm) thick. An experimental version of this aerotrain has reached more than 215 mph (346 km/h).

Hovercraft are sometimes called air-cushion vehicles, or ACVs, because they ride on a cushion of air. There are various ways of creating the air cushion, but on most hovercraft powerful fans force the air down under the craft, while a rubber skirt keeps it in place. By aiming the air downward and inward just inside the skirt, leakage is kept to a minimum. Most hovercraft are pushed forward by propellers on top, like an aircraft.

Hovercraft can move easily over any fairly flat surface and can skim from water to land without stopping. They are the only ground vehicles that can travel easily across marshland.

Some armies are hoping to use high-speed hovercraft to land troops. Because they can travel easily over marshes and other surfaces, hovercraft can zoom in to land troops on over 70 percent of the world's coasts. Conventional landing craft can only land troops on 15 percent.

Hovercraft have many other uses apart from military. In Canada, they are used for reaching far-flung outposts over land, water, and ice. In Europe, giant hovercraft are used as high-speed ferries. In future, people may travel in "flarecraft" —small hovercraft that swish over grass, marsh, and water at 150 mph (240 km/h).

In the real world

Christopher Cockerell's one-man SRN1 was the first successful hovercraft.

THE FIRST HOVERCRAFT

As long ago as 1716, Swedish philosopher Emmanuel Swedenborg came up with a design for a hovercraft, and in the 1870s, John Thorneycroft built model air-cushion craft. But the first to build a working hovercraft was British inventor Christopher Cockerell in the 1950s. In 1959, his famous SRN1 became the first hovercraft to cross the English Channel.

While hovercraft float over water on a layer of air, hydrofoils skim over the surface on narrow wings or "foils." The foils work like an airplane's wings to lift the craft out of the water, but because water is 1,000 times denser than air, the foils can be quite small. Hydrofoils can reach speeds of 70 mph (113 km/h). They are used as ferries in Europe and Asia. In the United States, they are used by the army.

MAKING A HOVERCRAFT

You will need

- ✔ An empty plastic water or soft drinks bottle (with its screw cap)
- ✔ A kitchen skewer to be used only by an adult
- ✔ Various round, good quality balloons
- ✔ Scissors

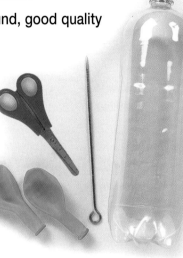

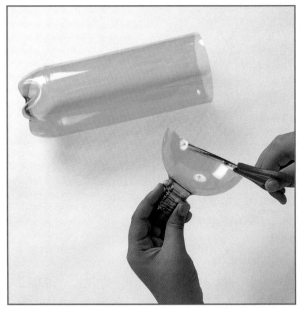

1 Ask an adult to skewer a hole in the cap of a soft drinks bottle, then cut the top off the bottle and trim the cut edge flat.

In the real world

RIDING ON AIR

Hovercraft ride on a thick cushion of air to speed across the ground. Many other things float on a thin film of air so they slide easily, just like the balloon hovercraft in this experiment. Heavy loads are often made to glide on a thin film of air. Special air bearings may also trap a layer of air between two surfaces to ensure they move together with minimum friction. Air bearings are often used in precision equipment such as astronomical telescopes. They ensure the telescope moves smoothly around to track the stars with total accuracy.

3 Still squeezing the neck of the balloon tightly, ease the balloon's opening carefully over the bottle cap.

You have created a simple air-cushion vehicle. Air rushing out from the balloon lifts it just off the table surface and keeps friction between the edge of the bottle top and the table to a minimum. The bottle top slides easily in any direction. This is why hovercraft can move so fast.

2 Inflate a balloon, then squeeze the neck tight to keep air from escaping.

Keep pinching the neck of the balloon to stop air escaping, then, with your other hand, screw the cut-off bottle top into the cap. Place the bottle top down on a smooth, flat table or floor. Let go of the balloon neck. The air should start to escape at once through the hole in the bottle cap. As it does, it will rush out under the edges of the bottle top, lifting the bottle top up very slightly. Try gently pushing the bottle top to and fro and you should find it slides about very easily. When the balloon has deflated, try pushing it again and you should find it much stickier.

SUBMARINES AND SUBMERSIBLES

Submarines, like the one below, have allowed us to explore the ocean and its treasures.

Unlike ships, submarines are designed not to float on the surface but move under the water. To do this, they still have to control their buoyancy. Indeed, it is by controlling their buoyancy that they are able to move up and down easily through the water.

Wrapped around the hull of a submarine are long, hollow cavities called ballast tanks. To dive below the surface, the submarine opens ballast tanks

In focus

DIVING AND SUBAQUA

Like submarines, divers move up and down through the water. It would be too exhausting to swim all the time to maintain depth. So divers have a special vest with an air bladder, or buoyancy compensator device (BCD). Adding air from the breathing tank to the BCD makes the diver more buoyant, allowing him or her to rise. Letting it out reduces buoyancy, so the diver sinks. By making small adjustments, the diver can achieve neutral buoyancy—neither floating up nor sinking down. Divers also wear a belt with lead weights to help them descend and stay underwater. A quick-release buckle on the belt allows the diver to shed the weights and make a rapid ascent to the surface in an emergency.

Divers aim to achieve neutral buoyancy by letting air in or out of the air bladder in their vests.

to allow seawater in. The extra weight of seawater reduces the submarine's buoyancy and lets it sink. By varying the number of tanks filled with water, a submarine can run at particular depths underwater.

To surface again, the submarine uses compressed air stored in cylinders. The air is let into the ballast tanks to blow water out so that the submarine becomes light and rises again.

While underwater, the submarine's propellers thrust it forward, while fins at both ends called hydroplanes tilt to drive it

up or down. Submarines cannot use diesel or petrol engines underwater, because there is no oxygen. They are driven either by electric motors from batteries or, in larger submarines, by turbines driven by nuclear reactors.

Submarines are the largest underwater craft. The USS *Seawolf* is over 350 ft (100 m) long and can travel around the world while submerged 1,500 ft (460 m) underwater. Smaller craft are called submersibles. Robot submersibles explore the deepest oceans.

FLOATING ROCKS

A rock thrown in a pool sinks instantly. Yet rocks can and do float if they are on denser material. In fact, all the world's rocks float on Earth's surface.

When Earth formed 4,600 million years ago, all the materials it is made of were mixed up in a hot mass. But as it cooled down, elements began to separate out. Heavy elements such as iron sank to Earth's center. Lighter elements such as oxygen and silicon drifted up to the surface. Some heavy elements also rose because they joined with oxygen to make substances called oxides, and with oxygen and silicon to make substances called silicates.

Eventually, after many millions of years of floating and sinking, Earth had clear layers, each made of different materials. In the center is a core of iron. At the surface is a thin layer of mainly oxides and silicates. In between is a very, very thick layer of mixed rock material called the mantle.

The surface layer hardened and cooled to form a thin, rigid crust. All the rocks and

Just as chunks of ice float on water, so Earth's crust— and the continents embedded in it —floats on Earth's partially melted mantle. This is a satellite view of the Bering Strait off Alaska.

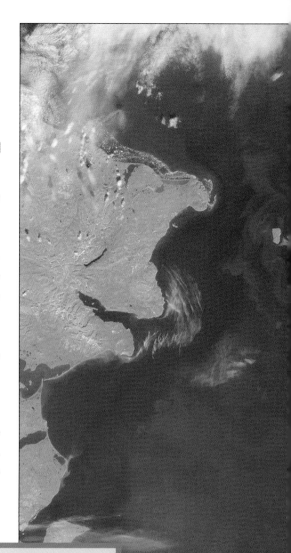

Did you know?

There are especially powerful convection currents in the earth's mantle called mantle plumes. Like giant fountains thousands of miles high, these plumes spout up in places beneath the crust, cracking it apart as they almost burst through. The shapes of the world's continents were made by cracks like these.

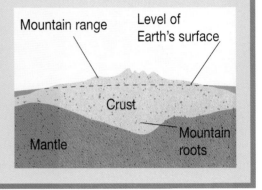

In focus

FLOATING MOUNTAINS

The pieces of Earth's crust float on the mantle just like a ship in the sea. Just as a ship floats lower in the water when heavily laden, so the crust floats lower in the mantle if it is heavy above. High mountain areas sink further, so have deep "roots" to balance their height. This is called isostasy. When mountains are worn down by the weather, rivers, and earthquakes, they float upward, maintaining the balance.

This diagram shows mountain roots sinking into the mantle.

Mountain range
Level of Earth's surface
Crust
Mantle
Mountain roots

continents you see on the surface are part of this crust.

Because the mantle is very hot, it is soft and flows very slowly like molasses. The lighter crust floats on top. The crust is actually broken into 20 or so fragments, or tectonic plates, that drift on the mantle like chunks of ice on a pond.

The world's continents are embedded in these plates and, as plates drift, so do continents. 250 million years ago, there was just one big landmass. Today's continents formed as this land broke up and the pieces drifted around the world, moving a fraction of an inch each year.

Geologists believe the slow dance of the continents is driven by circulations in the mantle. When a liquid is heated from below—like soup in a pan—the parts of the liquid next to the heat expand, become less dense, and so float upward. This is called convection. At the same time, cooler, denser regions sink. The sinking liquid is drawn in under the rising liquid, warms and starts to rise. This creates a steady rising stream, or convection current. Soon the entire liquid is circulating.

Geologists believe there are convection currents like these in Earth's mantle. Mantle material heated deep down by the hot core floats up through the mantle and hits the underside of the crust. It flows out sideways, cools, then sinks. The plates and continents are carried on the sideways flow.

FLOATING IN THE AIR

Clouds in the sky are not quite floating. They are made of droplets of water so tiny that they sink very slowly.

Air is a mixture of gases—mainly oxygen and nitrogen—that is thick enough to provide enough support for certain things to float in it. In order for something to float in air, however, it must be less dense than air.

As the Sun steams water off oceans, lakes, and rivers, for instance, the water floats up into the sky. While liquid water is 1,000 times as dense as air, water vapor is barely half as dense, so floats in air. It goes on rising until it gets into air so

Did you know?

Clouds only form if there are tiny aerosols in the air for the water to condense on. Scientists believe that as industrial pollution puts more aerosols into the air, clouds are becoming thicker and brighter.

cold that it condenses to form tiny drops of liquid water. These drops of liquid water form clouds in the sky.

Water is heavier than air, but the drops that make up clouds are so small they sink very slowly, and are swept along by the wind. But if the drops grow big—perhaps as more water vapor condenses—they become so heavy that they fall as rain.

Not only water drops. Tiny dust particles and other minute specks of material, called aerosols, float in the air. In heavily polluted air, there is a huge quantity of aerosols floating in the air, including a great deal of soot. Aerosols would be denser than air if they clumped together in a solid mass, but they float because they are in such fine particles.

All the world's wind and weather is driven by floating and sinking air. In some places, the Sun warms the ground and the ground warms the air. This warm air expands and becomes less dense than the air around it, so it floats up. In cooler places, the air sinks.

The whole atmosphere is on the move, stirred into motion by these circulating currents of rising and sinking air. Winds are simply currents of air moving from regions where the air is sinking to regions where the air is rising.

In the real world

Astronauts carrying out tasks outside orbiting spacecraft float freely in any direction.

FLOATING IN SPACE

For astronauts orbiting the earth in space, gravity seems to have no power. They become weightless and float freely—even more freely than in water. Space scientists describe this as "neutral buoyancy." Things that are neutrally buoyant neither float nor sink. In water, this is a balance between the downward pull of gravity and the buoyant force (upthrust) of the water. In space, astronauts (and everything else) are neutrally buoyant because the pull of gravity is balanced by the speed at which they are all hurtling around Earth. This is like a lift falling so fast that people inside float off the floor. Inside a spacecraft, weightlessness is relatively easy to cope with, but when the astronauts have to go outside it can be very difficult. To train astronauts, the American space agency sends them to the Neutral Buoyancy Laboratory (NBL). Here there is a swimming pool large enough to contain a space shuttle orbiter. The astronauts work around this in spacesuits designed to be neutrally buoyant in water.

Experiments in Science

Science is about knowledge: it is concerned with knowing and trying to understand the world around us. The word comes from the Latin word, *scire*, to know.

In the early 17th century, the great English thinker Francis Bacon suggested that the best way to learn about the world was not simply to think about it, but to go out and look for yourself—to make observations and try things out. Ever since then, scientists have tried to approach their work with a mixture of observation and experiment. Scientists insist that an idea or theory must be tested by observation and experiment before it is widely accepted.

All the experiments in this book have been tried before, and the theories behind them are widely accepted. But that is no reason why you should accept them. Once you have done all the experiments in this book, you will know the ideas are true not because we have told you so, but because you have seen for yourself.

All too often in science there is an external factor interfering with the result which the scientist just has not thought of. Sometimes this can make the experiment seem to work when it has not, as well as making it fail. One scientist conducted lots of demonstrations to show that a clever horse called Hans could count things and tap out the answer with his hoof. The horse was indeed clever, but later it was found that rather than counting, he was getting clues from tiny unconscious movements of the scientist's eyebrows.

This is why it is very important when conducting experiments to be as rigorous as you possibly can. The more casual you are, the more "eyebrow factors" you will let in. There will always be some things that you can not control. But the more precise you are, the less these are likely to affect the outcome.

What went wrong?

However careful you are, your experiments may not work. If so, you should try to find out where you went wrong. Then repeat the experiment until you are absolutely sure you are doing everything right. Scientists learn as much, if not more, from experiments that go wrong as those that succeed. In 1929, Alexander Fleming discovered the first antibiotic drug, penicillin, when he noticed that a bacteria culture he was growing for an experiment had gone moldy—and that the mold seemed to kill the bacteria. A poor scientist would probably have thrown the moldy culture away. A good scientist is one who looks for alternative explanations for unexpected results.

Glossary

Aerosol: Tiny particles so small and light they float in the air. The air contains huge amounts of aerosols of dust and salt.

Ballast: Extra weight put in the bottom of a ship to weigh it down and make it more stable. This may be extra cargo, or simply stones, gravel, or sand.

Buoyancy: The natural lift something that floats gets when it is immersed in water, or any other liquid or gas that is denser than it.

Convection: When something gets warm and expands it becomes less dense and so tends to float upward within its surroundings. Convection is this upward movement of warm pockets.

Density: The amount of matter that is packed into a certain volume, typically measured in kg per cubic meter or lb per cubic foot.

Friction: When two things rub together, the roughness of each surface tends to drag and stop them sliding past each other. Friction is the force that stops them sliding.

Gravity: Gravity is the force of attraction between every bit of matter in the universe. On Earth, gravity is the force of attraction between the earth and things on it, which makes them fall.

Hydrofoil: A wing that is added to a boat that creates lift in the water in just the same way that an airplane wing creates lift in the air.

Isostasy: The way all parts of the earth's crust float on the mantle so that the buoyant force that supports the crust is exactly balanced by the weight of the crust. This means that mountains, which are part of the crust, have deep roots so that there is enough buoyancy to support them.

Mantle: The thick middle layer of the earth beneath the crust and wrapped around the core. It is very hot and it flows very, very, very slowly.

Neutral buoyancy: A state in which the upthrust created by buoyancy exactly matches the weight of an object immersed in water. This means that it neither floats up nor sinks, but simply hangs in the same place.

Submarine: A large craft that can travel under the sea, usually a warship.

Submersible: A small submarine, typically used by engineers and scientists to explore the ocean. Many submersibles operate entirely automatically like robots and so have been able to explore the very deepest parts of the ocean.

Upthrust: The upward force created by buoyancy that pushes up all objects that are immersed in water or any other liquid or gas. It works in the opposite way to the object's weight.

Weightlessness: The apparent complete loss of weight that astronauts and orbiting spacecraft experience because they are falling around the earth.

Index